Cybersecurity

TODAY and TOMORROW

PAY
NOW
OR
PAY
LATER

Computer Science and Telecommunications Board

Division on Engineering and Physical Sciences

National Research Council

NATIONAL ACADEMY PRESS
Washington, D.C.

NATIONAL ACADEMY PRESS • 2101 Constitution Avenue, N.W. • Washington, D.C. 20418

NOTICE: The projects that are the basis of this synthesis report were approved by the Governing Board of the National Research Council, whose members are drawn from the councils of the National Academy of Sciences, the National Academy of Engineering, and the Institute of Medicine. The members of the committees responsible for the final reports of these projects and of the board that produced this synthesis were chosen for their special competences and with regard for appropriate balance.

Core support for the Computer Science and Telecommunications Board (CSTB) is provided by its public and private sponsors, which include federal agencies (the Air Force Office of Scientific Research, Defense Advanced Research Projects Agency, Department of Energy, National Aeronautics and Space Administration, National Institute of Standards and Technology, National Library of Medicine, National Science Foundation, and the Office of Naval Research); the Vadasz Family Foundation; and an evolving mix of charitable corporate and individual contributions. Sponsors enable but do not influence CSTB's work. Any opinions, findings, conclusions, or recommendations expressed in this publication are those of the authors and do not necessarily reflect the views of the organizations or agencies that provide support for CSTB.

International Standard Book Number 0-309-08312-5

Additional copies of this report are available from the Computer Science and Telecommunications Board, National Research Council, 2101 Constitution Avenue, N.W., Washington, DC 20418. Call 202-334-2605 or e-mail the CSTB at cstb@nas.edu. This report is also available online at <http://www.cstb.org>.

Printed in the United States of America

Suggested citation: Computer Science and Telecommunications Board, *Cybersecurity Today and Tomorrow: Pay Now or Pay Later*, National Academy Press, Washington, D.C., 2002.

THE NATIONAL ACADEMIES

National Academy of Sciences
National Academy of Engineering
Institute of Medicine
National Research Council

The **National Academy of Sciences** is a private, nonprofit, self-perpetuating society of distinguished scholars engaged in scientific and engineering research, dedicated to the furtherance of science and technology and to their use for the general welfare. Upon the authority of the charter granted to it by the Congress in 1863, the Academy has a mandate that requires it to advise the federal government on scientific and technical matters. Dr. Bruce M. Alberts is president of the National Academy of Sciences.

The **National Academy of Engineering** was established in 1964, under the charter of the National Academy of Sciences, as a parallel organization of outstanding engineers. It is autonomous in its administration and in the selection of its members, sharing with the National Academy of Sciences the responsibility for advising the federal government. The National Academy of Engineering also sponsors engineering programs aimed at meeting national needs, encourages education and research, and recognizes the superior achievements of engineers. Dr. Wm. A. Wulf is president of the National Academy of Engineering.

The **Institute of Medicine** was established in 1970 by the National Academy of Sciences to secure the services of eminent members of appropriate professions in the examination of policy matters pertaining to the health of the public. The Institute acts under the responsibility given to the National Academy of Sciences by its congressional charter to be an adviser to the federal government and, upon its own initiative, to identify issues of medical care, research, and education. Dr. Kenneth I. Shine is president of the Institute of Medicine.

The **National Research Council** was organized by the National Academy of Sciences in 1916 to associate the broad community of science and technology with the Academy's purposes of furthering knowledge and advising the federal government. Functioning in accordance with general policies determined by the Academy, the Council has become the principal operating agency of both the National Academy of Sciences and the National Academy of Engineering in providing services to the government, the public, and the scientific and engineering communities. The Council is administered jointly by both Academies and the Institute of Medicine. Dr. Bruce M. Alberts and Dr. Wm. A. Wulf are chairman and vice chairman, respectively, of the National Research Council.

iv

Preface

Starting with the publication of the report *Computers at Risk: Safe Computing in the Information Age* in 1991 (National Academy Press, Washington, D.C.), the Computer Science and Telecommunications Board (CSTB) has examined the issue of computer and communications security a number of times, from a number of perspectives. While there has been progress in security, it is a sad commentary on the state of the world that what CSTB wrote more than 10 years ago is still timely and relevant. For those who work in computer security, there is a deep frustration that research and recommendations do not seem to translate easily into deployment and utilization.

The events of September 11, 2001, suggest—indeed demand—that we take a renewed look at the security and robustness of our nation's infrastructure. Now, if ever, we see the importance of having critical systems resistant to attack and serviceable in times of crisis. From our telephone system to air traffic control to the Internet, we will be greatly harmed if these systems fail us just when we need them most.

The vulnerabilities are not new, only freshly brought into focus. And the approaches that will mitigate these threats are not unknown, only underutilized. So CSTB has taken the approach of drawing on its past work to point out that much of what we need to do is available to us now, if only we choose to act.

The staff of the CSTB have assembled this report from the broad base of its existing reports. Herb Lin deserves special thanks for the effort necessary to produce this report quickly.

David D. Clark, *Chair*
Computer Science and
Telecommunications Board

Acknowledgment of Reviewers

This report was reviewed in draft form by individuals chosen for their diverse perspectives and technical expertise, in accordance with procedures approved by the National Research Council's (NRC's) Report Review Committee. The purpose of this independent review is to provide candid and critical comments that will assist the institution in making the published report as sound as possible and to ensure that the report meets institutional standards for objectivity, evidence, and responsiveness to the study charge. The review comments and draft manuscript remain confidential to protect the integrity of the deliberative process. We wish to thank the following individuals for their participation in the review of this report:

Steven Bellovin, AT&T Labs Research,
Thomas Berson, Anagram Laboratories,
John Davis, Mitretek Systems Inc.,
Carl Landwehr, National Science Foundation,
Fred Schneider, Cornell University, and
Willis Ware, RAND Corporation.

Although the reviewers listed above have provided many constructive comments and suggestions, they were not asked to endorse the conclusions or recommendations, nor did they see the final draft of the report before its release. The review of this report was overseen by Gerry Dinneen. Appointed by the NRC's Report Review Committee, he was responsible for making certain that an independent examination of this report was carried out in accordance with institutional procedures and that all review comments were carefully considered. Responsibility for the final content of this report rests entirely with the Computer Science and Telecommunications Board and the National Research Council.

Contents

1   CYBERSECURITY TODAY AND TOMORROW   1
Background and Introduction, 1
The Nature of Cyberthreats, 2
Causes of System and Network Problems, 3
The Harm from Breaches of Cybersecurity, 6
What Do We Know About Cybersecurity?, 7
    General Observations, 7
    Management, 8
    Operational Considerations, 10
    Design and Architectural Considerations, 11
What Can Be Done?, 12
    Individual Organizations, 13
    Vendors of Computer Systems, 13
    Policy Makers, 14

2   EXCERPTS FROM EARLIER CSTB REPORTS   17
Computers at Risk: Safe Computing in the Information Age (1991), 18
    The Cybersecurity Challenge, 18
    Fundamentals of Cybersecurity, 18
    The Security Experience: Vulnerability, Threat, and
       Countermeasure, 20
    The Asymmetry Between Offense and Defense, 20
    Confidence in Countermeasures, 21

On Network Vulnerabilities, 21
Market Influences on Cybersecurity, 22
Nontechnical Dimensions of Cybersecurity, 22
Realizing the Potential of C4I: Fundamental Challenges (1999), 24
On What a Defense Must Do, 24
On Practice in the Field, 31
Trust in Cyberspace (1999), 33
Cybersecurity and Other Trustworthiness Qualities
Interact, 33
On Managing Risk, 33
Vulnerabilities in the Public Telephone Network
and the Internet, 35
On Building Secure Systems and Networks, 36
On the Impact of System Homogeneity ("Monoculture"), 37

WHAT IS CSTB?   39

1

Cybersecurity Today and Tomorrow

BACKGROUND AND INTRODUCTION

In the wake of the horrific events of September 11, 2001, the nation's attention has focused heavily on various dimensions of security. Security for nuclear power plants, shopping malls, sports stadiums, and airports, to name just a few entities, received additional scrutiny in the weeks that followed. Computer and telecommunication systems, too, have received significant attention, and because the Computer Science and Telecommunications Board (CSTB, described further on the final pages of this report) of the National Research Council (NRC) has examined various dimensions of computer and network security and vulnerability, it decided to revisit reports relevant to cybersecurity issued over the last decade. In some instances, security issues were the primary focus of a report from the start (see, for example, (1) *Computers at Risk*, 1991;[1] (2) *Cryptography's Role in Securing the Information Society*, 1996;[2] (3) *For the Record: Protecting Electronic Health Information*, 1997;[3] and (4) *Trust in Cyberspace*, 1999[4]). In

---

[1]Computer Science and Telecommunications Board, National Research Council. 1991. *Computers at Risk: Safe Computing in the Information Age*. National Academy Press, Washington, D.C.

[2]Kenneth W. Dam and Herbert S. Lin (eds.), Computer Science and Telecommunications Board, National Research Council. 1996. *Cryptography's Role in Securing the Information Society*. National Academy Press, Washington, D.C.

[3]Computer Science and Telecommunications Board, National Research Council. 1997. *For the Record: Protecting Electronic Health Information*. National Academy Press, Washington, D.C.

[4]Computer Science and Telecommunications Board, National Research Council. 1999. *Trust in Cyberspace*. National Academy Press, Washington, D.C.

other instances, security issues emerged as a prominent element of a study as the study unfolded (see, for example, (5) *Continued Review of the Tax Systems Modernization of the Internal Revenue Service*, 1996;[5] (6) *Realizing the Potential of C4I*, 1999;[6] and (7) *Embedded, Everywhere*, 2001[7]). (Hereinafter, these reports are referenced by number.)

Security issues continue to be an important part of CSTB's portfolio, and CSTB recently held workshops that explored how to deal with the insider threat to security (2000) and various legal issues associated with protecting critical infrastructure (2001). Though the most recent of the comprehensive reports was issued 2 years ago and the earliest 11 years ago, not much has changed with respect to security as it is practiced, notwithstanding further evolution of the public policy framework and an increase in our perception of the risks involved. The unfortunate reality is that relative to the magnitude of the threat, our ability and willingness to deal with threats have, on balance, changed for the worse (6), making many of the analyses, findings, and recommendations of these reports all the more relevant, timely, and applicable today. This document presents the enduring findings and recommendations from that body of work, and it includes excerpts from three of the reports listed above.

THE NATURE OF CYBERTHREATS

Much of modern life depends on computers and computer networks. For many people, the most visible interaction they have with computers is typing at the keyboard of the computer. But computers and networks are critical for key functions such as managing and operating nuclear power plants, dams, the electric power grid, the air traffic control system, and the financial infrastructure. Computers are also instrumental in the day-to-day operations of companies, organizations, and government. Companies large and small rely on computers to manage payroll, to track inventory and sales, and to perform research and development. Distribution of food and energy from producer to retail consumer relies on computers and networks at every stage. Nearly everyone in business or government

---

[5]Computer Science and Telecommunications Board, National Research Council. 1996. *Continued Review of the Tax Systems Modernization of the Internal Revenue Service: Final Report.* National Academy Press, Washington, D.C.

[6]Computer Science and Telecommunications Board, National Research Council. 1999. *Realizing the Potential of C4I: Fundamental Challenges.* National Academy Press, Washington, D.C.

[7]Computer Science and Telecommunications Board, National Research Council. 2001. *Embedded, Everywhere: A Research Agenda for Networked Systems of Embedded Computers.* National Academy Press, Washington, D.C.

relies on electronic communications—whether telephone, fax, e-mail, or instant messages—which are obviously enabled by computers. Many (perhaps even most) computer systems are networked in some fashion today (4), most visibly and commonly via the vast collection of globally interconnected computer networks known as the Internet. A more recent trend is toward embedding computing capability in all kinds of devices and environments and networking embedded systems into larger systems (7). These trends make many computing and communications systems critical infrastructure in themselves and components of other kinds of critical infrastructure, from energy to transportation systems.[8]

What can go wrong with a computer system or network?

- It can become unavailable or very slow (1,4,6). That is, using the system or network at all becomes impossible, or nearly so. The e-mail does not go through, or the computer simply freezes, with the result that somebody is unable to get his or her job done in a timely way, be it servicing a customer or reacting to a crisis.

- It can become corrupted, so that it does the wrong thing or gives wrong answers (1-7). For example, data stored on the computer may become different from what it should be, as would be the case if medical or financial records were improperly modified. Or, freight manifests might be altered so that the wrong material is shipped, an obvious problem for any rapid military deployment.

- It can become leaky (1-7). That is, someone who should not have access to some or all of the information available through the network obtains such access. For example, a spy who gains access to files stored in an intelligence agency information system may be able to view very sensitive data.

CAUSES OF SYSTEM AND NETWORK PROBLEMS

What can cause something to go wrong in a computer system or network? It is useful to distinguish between accidental causes and deliberate causes. In general, accidental causes are natural (e.g., a lightning surge that destroys a power supply in a network that causes part of the network to fail) or human but nondeliberate (e.g., an accidental program-

---

[8]The President's Commission on Critical Infrastructure Protection included under the rubric of "critical infrastructure" telecommunications, electric power systems, gas and oil production and storage, banking and finance, transportation, water supply systems, government services, and emergency services. See President's Commission on Critical Infrastructure Protection. 1997. *Critical Foundations*. Washington, D.C.

ming error that causes a computer to crash under certain circumstances, or the unintended cutting of a communications cable during excavation). Accidental causes figure prominently in many aspects of trustworthiness beside security, such as safety or reliability (1,3,4,7).

Deliberate problems are the result of conscious human choice. In the context of seeking to understand the laws of physics, Einstein once said that while nature may be subtle, it is not malicious. But in dealing with deliberate problems, one is faced with malicious intent. A malicious human may seek to hide his or her tracks, making it difficult to identify the nature of the problem caused (or even to identify that a problem has been caused).[9] A malicious human can, in principle, tailor actions to produce a desired effect beyond the damage to the actual system attacked—unlike an accidental problem whose effects are randomly determined. Security experts often refer to the efforts of these malicious people as "attacks."[10] A central challenge in responding to an information system attack is identifying who the attacker is and distinguishing whether the motive is mischief, terrorism, or attack on the nation. A related challenge is determining whether events that are distant in time or space are related—parts of a given attack (1,4,6).

Note also that an attacker—who seeks to cause damage deliberately—may be able to exploit a flaw accidentally introduced into a system. System design and/or implementation that is poor by accident can result in serious security problems that can be deliberately targeted in a penetration attempt by an attacker.[11]

There are many ways to cause problems deliberately. One way—which receives a great deal of attention in the media—is through an attack

---

[9]Tracing attacks is generally difficult, because serious attackers are likely to launder their connections to the target. That is, an attacker will compromise some intermediate targets whose vulnerabilities are easy to find and exploit, and use them to launch more serious attacks on the ultimate intended target. This, of course, is what has happened in a number of distributed denial-of-service attacks against Web servers of certain U.S. companies and government agencies, in which a number of computers flooded their targets with bogus requests for service, thus making them unavailable to provide service to legitimate users.

[10]"Attack" is a word that has seemed excessive to some (particularly because many attacks have been traced to individuals with motivation more akin to that of a joyrider than a state-supported, well-organized attacker). Recent events suggest that "attack" is increasingly appropriate, insofar as it is analogous to more familiar or conventional forms of attacks on resources of different kinds, in either the military or civilian sector.

[11]A particularly insidious "accidental" problem arises because of the fact that the precise software configuration on any operational system (including applications, device drivers, and system patches) has almost certainly not been tested for security—there are simply too many possible configurations to test more than a small fraction explicitly. As new applications and device drivers are installed over time, an operational system is more likely to exhibit additional vulnerabilities that an attacker might exploit.

that arrives "through the wires." The Internet makes it possible to mount such an attack remotely, anonymously, and on a large scale (4). One example of such "cyber-only" attacks are computer viruses that infect a user's computer, take some destructive action such as deleting files on the network or local hard drive, and propagate themselves further, such as by e-mailing copies of themselves. A second example is a distributed denial-of-service attack, described in footnote 9.

The damage that a cyber-only attack causes may not be immediately (or ever) apparent. A successful attack may lay a foundation for later attacks, be set to cause damage well after the initial penetration, or enable the clandestine transmission of sensitive information stored on the attacked system (1,4,6). For example, a number of recent incidents have compromised the computers of unsuspecting home computer users by implanting unauthorized code; these computers were subsequently used as launch points in a coordinated and distributed denial-of-service attack.

Finally, the fact that the Internet connects many of the world's computers implies that a cyber-only attack can be launched from locations around the world, routed through other countries (perhaps clandestinely and unknown to anyone in those other countries), and directed against any U.S. computer on the Internet. The availability of a plethora of launch points and routes for cyberattack greatly complicates the ability to stop an attack before it reaches the security barriers of the U.S. computers in question, as well as to identify its source. The Internet's various intentional or inadvertent links to other communications networks make them potentially vulnerable to worldwide attacks as well (1,2,4,6).

A cyber-only attack is only one way to cause problems in a computer system or network. Other ways include the following:

• The compromise of a trusted insider who can provide system or network access to outsiders or use his or her access for improper purposes (e.g., providing passwords that permit outsiders to gain entry) (1,3-6). This trusted insider may be recruited covertly by hostile parties, planted well in advance of any action associated with an actual attack (the so-called "sleeper" problem), or tricked into taking some action that breaches system security (e.g., tricked into disclosing a password or installing software that permits access by malicious outsiders).

• Physical destruction of some key element of the system or network, such as critical data centers or communications links (4,6,7). Examples of physical vulnerabilities are various backhoe incidents in which accidental cutting of fiber-optic cables (both primary and backup!) resulted in major network outages, and the severe damage to a Verizon central office in the World Trade Center attack on September 11, 2001.

It is useful to distinguish between three important concepts of cyber-security. A *vulnerability* is an error or a weakness in the design, imple-mentation, or operation of a system. A *threat* is an adversary that is motivated to exploit a system vulnerability and is capable of doing so. *Risk* refers to the likelihood that a vulnerability will be exploited, or that a threat may become harmful. In this lexicon, a system that allows com-puter viruses to replicate or unauthorized users to gain access exhibits vulnerabilities. The creator of the virus or the unauthorized user is the threat to the system. Operating a system with known vulnerabilities in the presence of possible threats entails some risk that harm or damage will result.

THE HARM FROM BREACHES OF CYBERSECURITY

How do potential cyberdisasters compare with disasters in the physi-cal world? As the catastrophic events of September 11, 2001, demon-strate, disasters in the physical world can involve massive loss of life and damage to physical infrastructure over a very short period of time. The damage from most cyberattacks is unlikely to be manifested in such a manner—although interference with medical information systems and devices could affect lives. If undertaken by themselves, cyberattacks could compromise systems and networks in ways that could render communi-cations and electric power distribution difficult or impossible, disrupt transportation and shipping, disable financial transactions, and result in the theft of large amounts of money (1,2,4). Economic and associated social harm is a likely consequence of a large-scale cyberattack that is successful. That harm would involve at least opportunity costs—inter-ruption of business, forgoing of various activities and associated benefits, and so on.

While such results would qualify on any scale as disastrous, addi-tional harm can come from the interactions of cyber- and physical sys-tems under attack that endanger human life directly and affect physical safety and well-being (3,4,7). In particular, a large-scale coordinated cyberattack could occur at the same time as an attack on the physical infrastructure. For example, a successful cyberattack launched on the air traffic control system in coordination with airliner hijackings could result in a much more catastrophic disaster scenario than was seen on Septem-ber 11, 2001. Or compromising communications channels during a physi-cal attack of that magnitude could prevent government officials from responding to the attack, coordinating emergency response efforts, or even knowing whether the attack was still ongoing.

WHAT DO WE KNOW ABOUT CYBERSECURITY?

With the above perspective in mind, here are some of the main messages about cybersecurity that emerge from a review of CSTB reports.

General Observations

• In the United States, information system vulnerabilities, from the standpoint of both operations and technology, are growing faster than the country's ability (and willingness) to respond (1,2,4,6,7).

• Security is expensive, not only in dollars but especially in interference with daily work (1,2,4-6). It has no value when there is no attack (or natural/accidental disruption in the system environment).[12] Consequently, people tend to use as little of it as they think they can get away with. Exhortations to be more careful may work for a short time, but operational security can be maintained only by systematic and independently conducted "red team" attacks *and* correction of the defects that they reveal. Moreover, there are no widely accepted metrics for characterizing security, so it is difficult for a decision maker to know how much security a certain investment buys or whether that investment is enough.

• The overall security of a system is only as strong as its weakest link (1-7). System security is a holistic problem, in which technological, managerial, organizational, regulatory, economic, and social aspects interact. Weaknesses in *any* of these aspects can be very damaging, since competent attackers seek out weak points in the security of a network or system.

• The best is the enemy of the good. Risk management is an essential element of any realistic strategy for dealing with security issues (2-6). Experience has demonstrated—sadly—that the quest for perfection is the enemy of concrete, actionable steps that can provide improved but not perfect security. It is true that given enough time and effort, almost any security system can be breached. But that does not diminish the value of steps that can increase the difficulty of breaching security.

• Security is a game of action and reaction (1,4,6). When old vulnerabilities are corrected, attackers look for new paths of attack. On the other hand, it takes time to find those new paths, and during that time the system is more secure.

• Systems have many potential points of vulnerability, and an attacker is free to choose any one of them. For example, as antivirus soft-

---

[12]More precisely, designs and architectures and implementation methods that can be used to make a system more secure are often the same as those that enhance system reliability and trustworthiness. However, specific security features often are not valuable against natural or accidental disruption.

ware came to protect against conventional viruses, virus writers exploited a new channel—the macro capabilities of word processors—that was never intended to provide the capability for implementing viruses. Thus, security must be approached on a system level rather than on a piecemeal basis (1-7).

• Because cyberattacks can be conducted without leaving publicly visible evidence (unlike, for example, a plane crash), it is easy to cover them up (1,3,4,6). Reporting of attempts, successful and unsuccessful, to breach security—the where, when, and how of attacks—is essential both for forensics (to determine who is responsible and whether incidents in different places are part of the same attack) and for prevention (to defend against future attacks). Researchers, developers, and operators need this information to redesign systems and procedures to avoid future incidents, and national security and law enforcement agencies need it to defend the nation as a whole. Organizations that are attacked prefer to conceal attacks, because publicity may undermine public confidence, disclose adverse information, and make managers look bad. Weighing these costs and benefits should be a public policy issue, but so far the commercial and face-saving concerns of targets have dominated, and there is no effective reporting. The airline industry might be a good model to copy (in the sense that both accidents and near-misses are reported), and the information-sharing problem is being explored in the context of critical infrastructure protection, a perspective that emerged in the late 1990s.

Management

• From an operational standpoint, cybersecurity today is far worse than what known best practices can provide (1-6). Even without any new security technologies, much better security would be possible today if technology producers, operators of critical systems, and users took appropriate steps. But new technologies and new operating procedures—which would require additional investment for research and development—could make things even better.

• Because a secure system doesn't allow users to do any more than an insecure system, system and network operators in the private sector spend only as much on security as they can justify on business grounds—and this may be much less than the nation needs as a whole (1,3-6). (The same is true of government agencies that must work within budget constraints, though the detailed cost-benefit calculus may be different.)

• Further, because serious cyberattacks are rare, the payoff from security investments is uncertain (and in many cases, it is society rather than any individual firm that will capture the benefit of improved security). As a result, system and network operators tend to underinvest in security. Changing market incentives—for example, by adjusting the liability to which business users of technology might be subject or the insurance implications of good security—could have a dramatic impact on the market for security features.[13]

• For economic reasons, systems are generally built out of commercial off-the-shelf components. These are not very secure because there isn't much market demand: Customers buy features and performance rather than security. The failure of the U.S. government's Orange Book[14] program even within the federal marketplace is a striking example. The government demanded secure systems, industry produced them, and then government agencies refused to buy them because they were slower and less functional than other nonsecure systems available on the open market (1,4).[15]

• Because security measures are disaster-preventing rather than pay-off-producing, a central aspect of security must be accountability. That is, users and operators must be held responsible by management for taking all appropriate security measures—one cannot count on financial and market incentives alone to drive appropriate action (1,3-6). Many security problems exist not because a fix is unknown but because some responsible party has not implemented a known fix. Of course, appropriate security measures are not free. Management must be willing to pay the costs and must demand from vendors the tools needed to minimize those costs. (Note that costs can include the costs of testing a fix to see if it ruins the production environment.) Management must resolve the conflict between holding people responsible and full reporting of problems, which tends to be easier in an environment in which individuals are not fearful of reporting problems.

---

[13]For example, under today's practices, a party that makes investments to prevent its own facilities from being used as part of a distributed denial-of-service (DDOS) attack will reap essentially no benefits from such investments, because such an attack is most likely to be launched against a different party. But today's Internet-using society would clearly benefit if many firms made such investments. Making parties liable for not securing their facilities against being illicitly used as part of a DDOS attack (today there is zero liability) would change the incentives for making such investments. In a current project on critical infrastructure protection and the law, CSTB is exploring this issue (among others) in greater depth.

[14]"The Orange Book" is the nickname for the Trusted Computer System Evaluation Criteria, which were intended to guide commercial system production generally and thereby improve the security of systems in use.

[15]It did not help that systems compliant with Orange Book criteria also came later to market and often had less functionality (e.g., in some cases, a certified system was unable to connect to a network, because a network connection was not part of the certified configuration).

Operational Considerations

• To promote accountability, frequent and unannounced penetration testing (so-called red-teaming) is essential to understand the actual operational vulnerabilities of deployed systems and networks (6). No other method is as effective at pointing to security problems that must be solved. Information about vulnerabilities thus gathered must be made available to those who are in a position to fix them—or to upper management, who can force them to be fixed. Note that effective red-teaming is undertaken independently of the system or network being tested—those being tested must not know when the test will occur or what aspects of security will be tested, while those doing the testing must be technically savvy and not constrained by operating orders that limit what they are permitted to do.

• Many compromises of an information system or network result from improper configuration (1,3,4,6). For example, a given system on a network may have a modem attached to it that is not known to the network administrator, even if it was attached by a legitimate user for legitimate purposes. An installed operating system on a computer may lack critical "bug" fixes, because they were not applied or because the system was restored from a backup tape that did not include those fixes. A system firewall may be improperly configured in a way that allows Web access when, in fact, the system should only be able to transmit and receive e-mail. Or, a group of users may be given privileges that should, in fact, be restricted to one member of that group. Because checking operational configurations is very labor-intensive if done manually, it is essential to have configuration management tools for both systems and networks that can automatically enforce a desired configuration or alert administrators when variances from the known configuration are detected. Such tools are miserably inadequate today. Building them does not require research, but it does require a considerable amount of careful engineering.

• Since perfect security is impossible, secure configurations need to be updated when new attacks are discovered. These updates need to be delivered automatically to millions of systems (4,7).[16]

• Organizations must have concrete fallback action plans that instruct users and administrators about what they should do under condi-

---

[16]On the other hand, there is a nontrivial chance that updates will diminish existing and needed functionality, and people are sometimes reluctant to apply updates because they are reluctant to instigate system instability. Thus, the trustworthiness of the updates themselves, as well as the updating process, becomes an issue of concern.

tions of cyberattack (6). Admonitions to "be more careful" are not action-able, especially since the effects of a cyberattack may not be obvious, nor will they be constant over time. Instead, the tradeoffs between vulner-ability and functionality must be understood, and appropriate responses to attacks must be defined. Usually these responses involve making the systems do less in order to make them less vulnerable: fewer authorized users, less software running, and less communication between systems. For example, software architects might design a system so that operators could close off certain routes of access to it when under attack, thereby losing the useful functionality associated with those routes but preserv-ing critical functions that do not need those routes.

Design and Architectural Considerations

- There are often tensions between security and other good things, such as features, ease of use, and interoperability (1,3,4,6). For example, if users have different passwords for different systems, it is harder for an unauthorized party to gain access to all of those systems, but users must bear the burden of remembering multiple passwords. Because the ben-efits of successful security can be seen only in events that usually do not happen, resources devoted to security are "wasted" in the same sense that resources devoted to insurance are "wasted." In both cases, the system user (or the insured party) does not gain additional functionality as the result of its expenditures. But that does not make investments in security worthless—rather, it changes the terms on which such investments should be evaluated, which include the value of being able to continue operation in the face of hostile attacks (and often natural or accidental disruptions as well).[17]
- Human error is usually not a useful explanation for security prob-lems. Usually either operational or management practice is at fault: op-erational practice that requires people to get too many details right or that does too little red-team testing, and management practice that allows too little time for security procedures or fails to ensure that problems uncov-ered by testing are fixed (1,3-5).
- While cryptography is *not* a magic bullet for security problems, it

---

[17]Note that one fundamental difference between risks in the physical world and risks in cyberspace is the existence of an extensive actuarial database for the former that enables organizations to assess the payoff from investments to deal with those risks. By compari-son, operations in cyberspace are new and continually evolving, and risks in cyberspace are not well understood by the insurance industry. That industry has recently increased its activity in this domain, but progress has been slow.

does play key roles in system and network security. Cryptography has three primary uses: authentication, integrity checks, and confidentiality (1,2,4). Cryptographic authentication is an important aspect of security techniques that deny access or system privileges to unauthorized users. Cryptographic integrity checks ensure that data cannot be modified without revealing the fact of modification. Cryptographic confidentiality can be used to keep unauthorized parties from reading data stored in systems and sent over networks. Of course, adversaries can also use cryptography, to the detriment of certain national security and law enforcement purposes, but its widespread use can promote and enhance crime prevention and national security efforts as well.

• User authentication is essential for access control and for auditing (1-6). The most common method used today to authenticate users is passwords, which are known to be insecure compared with other authentication methods. A hardware token (e.g., smart card), supplemented by a personal identification number or biometrics (assuming good implementation), is much better. The user doesn't have to keep track of passwords, and a lost token is physically obvious and cannot be broadcast to a myriad of unauthorized parties (but the user does have to remember to bring the token to the computer access point).

• A common approach to network security is to surround an insecure network with a defensive perimeter that controls access to the network (1,4,6). Once past the perimeter, a user is left unconstrained. A perimeter defense is good as *part* of a defense in depth, especially because the security burden is placed primarily on those who manage the perimeter rather than those who manage systems inside the perimeter. However, it is entirely vulnerable if a hostile party gains access to a system inside the perimeter or compromises a single authorized user. Another approach to network security is mutual suspicion: Every system within a critical network regards every other system as a potential source of threat. Thus, a hostile party who gains access to one system does not automatically gain access to the whole network. Mutual suspicion can provide significantly higher levels of security, but it requires all system operators to pay attention to security rather than just those at the network perimeter. Perimeter defense and mutual suspicion can be used together to increase network security.

WHAT CAN BE DONE?

It is helpful to distinguish among actions that can be taken by individual organizations, by vendors, and by makers of public policy. However, the best results for cybersecurity will be obtained through actions by all parties.

Individual Organizations

Individual organizations should:

• Establish and provide adequate resources to an internal entity with responsibility for providing direct defensive operational support to system administrators throughout the organization (3,5,6). To serve as the focal point for operational change, such an entity must have the authority—as well as a person in charge—to force corrective action.
• Ensure that adequate information security tools are available, that everyone is properly trained in their use, and that enough time is available to use them properly. Then hold all personnel accountable for their information system security practices (3,5,6).
• Conduct frequent, unannounced red-team penetration testing of deployed systems and report the results to responsible management (6).
• Promptly fix problems and vulnerabilities that are known or that are discovered to exist (3,5,6).
• Mandate the organization-wide use of currently available network/configuration management tools, and demand better tools from vendors (3,5,6).
• Mandate the use of strong authentication mechanisms to protect sensitive or critical information and systems (3,5,6).
• Use defense in depth. In particular, design systems under the assumption that they will be connected to a compromised network or a network that is under attack, and practice operating these systems under this assumption (1,4,6).
• Define a fallback plan for more secure operation when under attack and rehearse it regularly. Complement that plan with a disaster recovery program (1,6).

Vendors of Computer Systems

Vendors of computer systems should:

• Drastically improve the user interface to security, which is totally incomprehensible in nearly all of today's systems (1,4,6). Users and administrators must be able to easily see the current security state of their systems; this means that the state must be expressible in simple terms.
• Develop tools to monitor systems automatically for consistency with defined secure configurations, and enforce these configurations (1, 4,6,7). Extensive automation is essential to reduce the amount of human labor that goes into security. The tools must promptly and automatically respond to changes that result from new attacks.

• Provide well-engineered schemes for user authentication based on hardware tokens (3,4,6). These systems should be both more secure and more convenient for users than are current password systems.

• Develop a few simple and clear blueprints for secure operation that users can follow, since most organizations lack the expertise to do this properly on their own. For example, systems should be shipped with security features turned on, so that a conscious effort is needed to disable them, and with default identifications and passwords turned off, so that a conscious effort is needed to select them (1,3).

• Strengthen software development processes and conduct more rigorous testing of software and systems for security flaws, doing so before releasing products rather than use customers as implicit beta testers to shake out security flaws (4).[18] Changing this mind-set is one necessary element of an improved cybersecurity posture.

Policy Makers

Policy makers should:

• Consider legislative responses to the failure of existing incentives to cause the market to respond adequately to the security challenge. Possible options include steps that would increase the exposure of software and system vendors and system operators to liability for system breaches and mandated reporting of security breaches that could threaten critical societal functions (1).

• Position the federal government as a leader in technology use and practice by requiring agencies to adhere to the practices recommended above and to report on their progress in implementing those measures (1,2,5).[19] Such a step would also help to grow the market for security technology, training, and other services.

• Provide adequate support for research and development on information systems security (1,4,7). Research and development on informa-

---

[18]Note that security-specific testing of software goes beyond looking at flaws that emerge in the course of ordinary usage in an Internet-connected production environment. For example, security-specific testing may involve very sophisticated attacks that are not widely known in the broader Internet hacker community.

[19]This concept has been implicit in a series of laws, beginning with the Computer Security Act of 1987, and administrative guidance (e.g., from the Office of Management and Budget and more recently from the Federal Chief Information Officers Council). Although it has been an elusive goal, movements toward e-government have provided practical, legal, and administrative impetus.

tion systems security should be construed broadly to include R&D on defensive technology (including both underlying technologies and architectural issues), organizational and sociological dimensions of such security, forensic and recovery tools, and best policies and practices. Given the failure of the market to address security challenges adequately, government support for such research is especially important.

2

Excerpts from Earlier CSTB Reports

This chapter contains excerpts from three CSTB reports: *Computers at Risk* (1991), *Realizing the Potential of C4I* (1999), and *Trust in Cyberspace* (1999). While this synthesis report is based on all of the references in footnotes 1-7 (Chapter 1), the excerpts from these three reports are the most general and broad. To keep this report to a reasonable length, nothing was excerpted from the other four reports. Readers are encouraged to read all of these reports, which can be found online at <http://www.nap.edu>. For the sake of simplicity and organizational clarity, footnotes appearing in the original text have been omitted from the reprinted material that follows. A gray bar in the margin, rather than indentation, is used to indicate extracted text. Subsection heads have been added to show the topics addressed.

COMPUTERS AT RISK:
SAFE COMPUTING IN THE INFORMATION AGE (1991)

CITATION: Computer Science and Telecommunications Board (CSTB), National Research Council. 1991. *Computers at Risk: Safe Computing in the Information Age.* National Academy Press, Washington, D.C.

The Cybersecurity Challenge

(From pp. 7-8): We are at risk. Increasingly, America depends on computers. They control power delivery, communications, aviation, and financial services. They are used to store vital information, from medical records to business plans to criminal records. Although we trust them, they are vulnerable—to the effects of poor design and insufficient quality control, to accident, and perhaps most alarmingly, to deliberate attack. The modern thief can steal more with a computer than with a gun. Tomorrow's terrorist may be able to do more damage with a keyboard than with a bomb.

To date, we have been remarkably lucky. Yes, there has been theft of money and information, although how much has been stolen is impossible to know. Yes, lives have been lost because of computer errors. Yes, computer failures have disrupted communication and financial systems. But, as far as we can tell, there has been no successful systematic attempt to subvert any of our critical computing systems. Unfortunately, there is reason to believe that our luck will soon run out. Thus far we have relied on the absence of malicious people who are both capable and motivated. We can no longer do so. We must instead attempt to build computer systems that are secure and trustworthy.

. . .[T]he degree to which a computer system and the information it holds can be protected and preserved . . . , which is referred to here as computer security, is a broad concept; security can be compromised by bad system design, imperfect implementation, weak administration of procedures, or through accidents, which can facilitate attacks. Of course, if we are to trust our systems, they must survive accidents as well as attack. Security supports overall trustworthiness, and vice versa.

Fundamentals of Cybersecurity

(From p. 2): Security refers to protection against unwanted disclosure, modification, or destruction of data in a system and also to the safeguarding of systems themselves. Security, safety, and reliability together are elements of system trustworthiness—which inspires the confidence that a system will do what it is expected to do.

(From pp. 49-50): Organizations and people that use computers can describe their needs for information security and trust in systems in terms of three major requirements:

- *Confidentiality:* controlling who gets to read information;
- *Integrity:* assuring that information and programs are changed only in a specified and authorized manner; and
- *Availability:* assuring that authorized users have continued access to information and resources.

These three requirements may be emphasized differently in various applications. For a national defense system, the chief concern may be ensuring the confidentiality of classified information, whereas a funds transfer system may require strong integrity controls. The requirements for applications that are connected to external systems will differ from those for applications without such interconnection. Thus the specific requirements and controls for information security can vary.

The framework within which an organization strives to meet its needs for information security is codified as security policy. A security policy is a concise statement, by those responsible for a system (e.g., senior management), of information values, protection responsibilities, and organizational commitment. One can implement that policy by taking specific actions guided by management control principles and utilizing specific security standards, procedures, and mechanisms. Conversely, the selection of standards, procedures, and mechanisms should be guided by policy to be most effective.

To be useful, a security policy must not only state the security need (e.g., for confidentiality—that data shall be disclosed only to authorized individuals), but also address the range of circumstances under which that need must be met and the associated operating standards. Without this second part, a security policy is so general as to be useless (although the second part may be realized through procedures and standards set to implement the policy). In any particular circumstance, some threats are more probable than others, and a prudent policy setter must assess the threats, assign a level of concern to each, and state a policy in terms of which threats are to be resisted. For example, until recently most policies for security did not require that security needs be met in the face of a virus attack, because that form of attack was uncommon and not widely understood. As viruses have escalated from a hypothetical to a commonplace threat, it has become necessary to rethink such policies in regard to methods of distribution and acquisition of software. Implicit in this process is management's choice of a level of residual risk that it will live with, a level that varies among organizations.

The Security Experience: Vulnerability, Threat, and Countermeasure

(From pp. 13-14): The field of security has its own language and mode of thought, which focus on the processes of attack and on preventing, detecting, and recovering from attacks. In practice, similar thinking is accorded to the possibility of accidents that, like attacks, could result in disclosure, modification, or destruction of information or systems or a delay in system use. Security is traditionally discussed in terms of vulnerabilities, threats, and countermeasures. A vulnerability is an aspect of some system that leaves it open to attack. A threat is a hostile party with the potential to exploit that vulnerability and cause damage. A countermeasure or safeguard is an added step or improved design that eliminates the vulnerability and renders the threat impotent.

A safe containing valuables, for example, may have a noisy combination lock—a vulnerability—whose clicking can be recorded and analyzed to recover the combination. It is surmised that safecrackers can make contact with experts in illegal eavesdropping—a threat. A policy is therefore instituted that recordings of random clicking must be played at loud volume when the safe is opened—a countermeasure.

Threats and countermeasures interact in intricate and often counterintuitive ways: a threat leads to a countermeasure, and the countermeasure spawns a new threat. Few countermeasures are so effective that they actually eliminate a threat. New means of attack are devised (e.g., computerized signal processing to separate "live" clicks from recorded ones), and the result is a more sophisticated threat.

The Asymmetry Between Offense and Defense

(From p. 14): The interaction of threat and countermeasure poses distinctive problems for security specialists: the attacker must find but one of possibly multiple vulnerabilities in order to succeed; the security specialist must develop countermeasures for all. The advantage is therefore heavily to the attacker until very late in the mutual evolution of threat and countermeasure.

If one waits until a threat is manifest through a successful attack, then significant damage can be done before an effective countermeasure can be developed and deployed. Therefore countermeasure engineering must be based on speculation. Effort may be expended in countering attacks that are never attempted. The need to speculate and to budget resources for countermeasures also implies a need to understand what it is that should be protected, and why; such understanding should drive the choice of a protection strategy and countermeasures. This thinking should be captured in security policies generated by management; poor security often reflects both weak policy and inadequate forethought.

Confidence in Countermeasures

(From p. 15): Confidence in countermeasures is generally achieved by submitting them for evaluation by an independent team; this process increases the lead times and costs of producing secure systems. The existence of a successful attack can be demonstrated by an experiment, but the adequacy of a set of countermeasures cannot. Security specialists must resort to analysis, yet mathematical proofs in the face of constantly changing systems are impossible.

In practice, the effectiveness of a countermeasure often depends on how it is used; the best safe in the world is worthless if no one remembers to close the door. The possibility of legitimate users being hoodwinked into doing what an attacker cannot do for himself cautions against placing too much faith in purely technological countermeasures.

The evolution of countermeasures is a dynamic process. Security requires ongoing attention and planning, because yesterday's safeguards may not be effective tomorrow, or even today.

On Network Vulnerabilities

(From p. 17): Interconnection gives an almost ecological flavor to security; it creates dependencies that can harm as well as benefit the community of those who are interconnected. An analogy can be made to pollution: the pollution generated as a byproduct of legitimate activity causes damage external to the polluter. A recognized public interest in eliminating the damage may compel the installation of pollution control equipment for the benefit of the community, although the installation may not be justified by the narrow self-interest of the polluter. Just as average citizens have only a limited technical understanding of their vulnerability to pollution, so also individuals and organizations today have little understanding of the extent to which their computer systems are put at risk by those systems to which they are connected, or vice versa. The public interest in the safety of networks may require some assurances about the quality of security as a prerequisite for some kinds of network connection.

(From p. 8): The threats to U.S. computer systems are international, and sometimes also political. The international nature of military and intelligence threats has always been recognized and addressed by the U.S. government. But a broader international threat to U.S. information resources is emerging with the proliferation of international computer networking—involving systems for researchers, companies, and other organizations and individuals—and a shift from conventional military conflict to economic competition. The concentration of information and economic activity in computer systems makes those systems an attractive target to hostile entities. This prospect raises questions about the intersection of economic and national security interests and the design of appropriate security strategies for the public and private sectors. Finally, politically motivated attacks may also target a new class of system that is neither commercial nor military: computerized voting systems.

Market Influences on Cybersecurity

(From pp. 159-161): Even the best product will not be sold if the consumer does not see a need for it. Consumer awareness and willingness to pay are limited because people simply do not know enough about the likelihood or the consequences of attacks on computer systems or about more benign factors that can result in system failure or compromise. Consumer appreciation of system quality focuses on features that affect normal operations—speed, ease of use, functionality, and so on. This situation feeds a market for inappropriate or incomplete security solutions, such as antiviral software that is effective only against certain viruses but may be believed to provide broader protection, or password identification systems that are easily subverted in ordinary use. . . .

Enhancing security requires changes in attitudes and behavior that are difficult because most people consider computer security to be abstract and concerned more with hypothetical rather than likely events. Very few individuals not professionally concerned with security, from top management through the lowest-level employee, have ever been directly involved in or affected by a computer security incident. Such incidents are reported infrequently, and then often in specialized media, and they are comprehensible only in broadest outline. Further, most people have difficulty relating to the intricacies of malicious computer actions. Yet it is understood that installing computer security safeguards has negative aspects such as added cost, diminished performance (e.g., slower response times), inconvenience in use, and the awkwardness of monitoring and enforcement, not to mention objections from the work force to any of the above. The Internet worm experience showed that even individuals and organizations that understand the threats may not act to protect against them.

Nontechnical Dimensions of Cybersecurity

(From p. 17): Computer security does not stop or start at the computer. It is not a single feature, like memory size, nor can it be guaranteed by a single feature or even a set of features. It comprises at a minimum computer hardware, software, networks, and other equipment to which the computers are connected, facilities in which the computer is housed, and persons who use or otherwise come into contact with the computer. Serious security exposures may result from any weak technical or human link in the entire complex. For this reason, security is only partly a technical problem: it has significant procedural, administrative, physical facility, and personnel components as well.

(From pp. 50-51): Technical measures alone cannot prevent violations of the trust people place in individuals, violations that have been the source of much of the computer security problem in industry to date . . . . Technical measures may prevent people from doing unauthorized things but cannot prevent them from doing things that their job functions entitle them to do. Thus, to prevent violations of trust rather than just repair the damage that results, one must depend primarily on human awareness of what other human beings in an organization are doing. But even a technically sound system with informed and watchful management and users cannot be free of all possible vulnerabilities. The residual risk must be managed by auditing, backup, and recovery procedures supported by general alertness and creative responses. Moreover, an organization must have administrative procedures in place to bring peculiar actions to the attention of someone who can legitimately inquire into the appropriateness of such actions, and that person must actually make the inquiry. In many organizations, these administrative provisions are far less satisfactory than are the technical provisions for security.

(From p. 10): It is important to balance technical and nontechnical approaches to enhancing system security and trust. Accordingly, the committee is concerned that the development of legislation and case law is being outpaced by the growth of technology and changes in our society. In particular, although law can be used to encourage good practice, it is difficult to match law to the circumstances of computer system use. Nevertheless, attacks on computer and communication systems are coming to be seen as punishable and often criminal acts . . . within countries, and there is a movement toward international coordination of investigation and prosecution. However, there is by no means a consensus about what uses of computers are legitimate and socially acceptable. Free speech questions have been raised in connection with recent criminal investigations into dissemination of certain computer-related information. There are also controversies surrounding the privacy impacts of new and proposed computer systems, including some proposed security safeguards. Disagreement on these fundamental questions exists not only within society at large but also within the community of computer specialists.

REALIZING THE POTENTIAL OF C4I:
FUNDAMENTAL CHALLENGES (1999)

CITATION: Computer Science and Telecommunications Board (CSTB), National Research Council. 1999. *Realizing the Potential of C4I: Fundamental Challenges.* National Academy Press, Washington, D.C.

C4I is a Department of Defense (DOD) acronym for command, control, communications, computers, and intelligence. While many of the information systems described in *Realizing the Potential of C4I: Fundamental Challenges* are owned or operated by the Department of Defense, essentially all of the implications and lessons for DOD systems are valid for non-DOD government systems, and for systems in the private sector.[20] Furthermore, the description of DOD practices in the field should not be taken as an exoneration of practices in non-DOD systems—indeed, it is highly likely that the same observations would apply to most such systems.

On What a Defense Must Do

(From pp. 144-152): Effective information systems security is based on a number of functions described below. This list of functions is not complete; nevertheless, evidence that all these functions are being performed in an effective and coordinated fashion will be evidence that information systems security is being taken seriously and conducted effectively.

---

[20]In 1991, *Computers at Risk* (at pp. 18-19) cast this point in the following terms:

There has been much debate about the difference between military and commercial needs in the security area. . . . This distinction is both superficial and misleading. National security activities, such as military operations, rely heavily on the integrity of data in such contexts as intelligence reports, targeting information, and command and control systems, as well as in more mundane applications such as payroll systems. Private sector organizations are concerned about protecting the confidentiality of merger and divestiture plans, personnel data, trade secrets, sales and marketing data and plans, and so on. Thus there are many common needs in the defense and civilian worlds.

Commonalities are especially strong when one compares the military to what could be called infrastructural industries—banking, the telephone system, power generation and distribution, airline scheduling and maintenance, and securities and commodities exchanges. Such industries both rely on computers and have strong security programs because of the linkage between security and reliability. Nonsecure systems are also potentially unreliable systems, and unreliability is anathema to infrastructure.

Some of these functions were also noted in the military context by the Defense Science Board, and some by the President's Commission on Critical Infrastructure Protection in its report. These functions are listed here because they are important, and because the committee believes that they have not yet been addressed by the DOD in an effective fashion (as described in the committee's findings below).

Function 1. Collect, analyze, and disseminate strategic intelligence about threats to systems.

Any good defense attempts to learn as much as possible about the threats that it may face, both the tools that an adversary may use and the identity and motivations of likely attackers. In the information systems security world, it is difficult to collect information about attackers (though such intelligence information should be sought). It is, however, much easier to collect and analyze information on technical and procedural vulnerabilities, to characterize both the nature of these vulnerabilities and their frequency at different installations. Dissemination of information about these vulnerabilities enables administrators of the information systems that may be affected to take remedial action.

Function 2. Monitor indications and warnings.

All defenses—physical and cyber—rely to some extent on indications and warning of impending attack. The reason is that if it is known that attack is impending, the defense can take actions to reduce its vulnerability and to increase the effectiveness of its response. This function calls for:

• *Monitoring of threat indicators.* For example, near-simultaneous penetration attempts on hundreds of military information systems might reasonably be considered an indication of an orchestrated attack. Mobilization of a foreign nation's key personnel known to have responsibility for information attacks might be another indicator. The notion of an "information condition" or "INFOCON," analogous to the defense condition (DEFCON) indicator, would be a useful summary device to indicate to commanders the state of the cyber-threat at any given time. This concept is being developed by various DOD elements but is yet immature.

• *Assessment and characterization of the information attack (if any).* Knowledge of the techniques used in an attack on one information system may facilitate a judgment of the seriousness of the attack. For example, an attack that involves techniques that are not widely known may indicate that the perpetrators have a high degree of technical sophistication.

- *Dissemination of information about the target(s) of threat.* Knowledge of the techniques used in an attack on one information system may enable administrators responsible for other systems to take preventive actions tailored to that type of attack. This is true even if the first attack is unsuccessful, because security features that may have thwarted the first attack may not necessarily be installed or operational on other systems.

Note that dissemination of information about attacks and their targets is required on two distinct time scales. The first time scale is seconds or minutes after the attack is known; such knowledge enables operators of other systems not (yet) under attack to take immediate preventive action (such as severing some network connections). In this instance, alternative means of secure communication may be necessary to disseminate such information. The second time scale is days after the attack is understood; such knowledge allows operators throughout the entire system of systems to implement fixes and patches that they may not yet have fixed, and to request fixes that are needed but not yet developed. . . .

Function 3. Be able to identify intruders.

Electronic intruders into a system are admittedly hard to identify. Attacks are conducted remotely, and a chain of linkages from the attacker's system to an intermediate node to another and another to the attacked system can easily obscure the identity of the intruder. Nevertheless, certain types of information—if collected—may shed some light on the intruder's identity. For example, some attackers may preferentially use certain tools or techniques (e.g., the same dictionary to test for passwords), or use certain sites to gain access. Attacks that go on over an extended period of time may provide further opportunities to trace the origin of the attack.

Function 4. Test for security weaknesses in fielded and operational systems.

An essential part of a security program is searching for technical and operational or procedural vulnerabilities. Ongoing tests (conducted by groups often known as "red teams" or "tiger teams") are essential for several reasons:

- Recognized vulnerabilities are not always corrected, and known fixes are frequently found not to have been applied as a result of poor configuration management.
- Security features are often turned off in an effort to improve operational efficiency. Such actions may improve operational efficiency, but at

the potentially high cost of compromising security, sometimes with the primary damage occurring in some distant part of the system.

- Some security measures rely on procedural measures and thus depend on proper training and ongoing vigilance on the part of commanders and system managers.
- Security flaws that are not apparent to the defender undergoing an inspection may be uncovered by a committed attacker (as they would be uncovered in an actual attack).

Thus, it is essential to use available tools and conduct red team or tiger team probes often and without warning to test security defenses. In order to maximize the impact of these tests, reports should be disseminated widely. Release of such information may risk embarrassment of certain parties or possible release of information that can be used by adversaries to attack, but especially in the case of vulnerabilities uncovered for which fixes are available, the benefits of releasing such information—allowing others to learn from it and motivating fixes to be installed—outweigh these costs.

Tiger team attacks launched without the knowledge of the attacked systems also allow estimates to be made of the frequency of attacks. Specifically, the fraction of tiger team attacks that are detected is a reasonable estimate of the fraction of adversary attacks that are made. Thus, the frequency of adversary attacks can be estimated from the number of adversary attacks that are detected.

Function 5. Plan a range of responses.

Any organization relying on information systems should have a number of routine information systems security activities (e.g., security features that are turned on, security procedures that are followed). But when attack is imminent (or in process), an organization could prudently adopt additional security measures that during times of non-attack might not be in effect because of their negative impact on operations. Tailoring in advance a range of information systems security actions to be taken under different threat conditions would help an organization plan its response to any given attack.

For example, a common response under attack is to drop non-essential functions from a system connected to the network so as to reduce the number of routes for penetration. A determination in advance of what functions count as non-essential and under what circumstances such a determination is valid would help facilitate an orderly transition to different threat conditions, and would be much better than an approach that calls for dropping all functionality and restoring only those functions that people using the system at the time complain about losing. Note that such determinations can be made only from an operational perspective rather than a technical one, a fact that points to the essential need for an operational architecture in the design of C4I systems.

The principle underlying response planning should be that of "graceful degradation"; that is, the system or network should lose functionality gradually, as a function of the severity of the attack compared to its ability to defend against it. This principle stands in contrast to a different principle that might call for the maintenance of all functionality until the attack simply overwhelms the defense and the system or network collapses. The latter principle is tempting because reductions in functionality necessitated for security reasons may interfere with operational ease of use, but its adoption risks catastrophic failure.

It is particularly important to note that designing a system for graceful degradation depends on system architects who take into account the needs of security (and more generally, the needs of coping with possible component failures) from the start. For example, the principle of graceful degradation would forbid a system whose continued operation depended entirely on a single component remaining functional, or on the absence of a security threat.

This principle is often violated in the development of prototypes. It is often said that "it is necessary for one to crawl before one can run," i.e., that it is acceptable to ignore security or reliability considerations when one is attempting to demonstrate the feasibility of a particular concept. This argument is superficially plausible, but in practice it does not hold water. It is reasonable for a prototype to focus only on concept feasibility, ignoring considerations of reliability or security, only if the prototype will be thrown away and a new architecture is designed and developed from scratch to implement the concept. Budget and schedule constraints usually prevent such new beginnings, and so in practice the prototype's architecture is never abandoned, and security or reliability considerations must be addressed in the face of an architecture that was never designed or intended to support them.

Function 6. Coordinate defensive activities throughout the enterprise.

Any large, distributed organization has many information systems and subnetworks that must be defended. The activities taken to defend each of these systems and networks must be coordinated because the distributed parts have interconnections and the security of the whole organization depends on the weakest link. Furthermore, it is important for different parts of organizations to be able to learn from each other about vulnerabilities, threats, and effective countermeasures.

Function 7. Ensure the adequacy, availability, and functioning of public infrastructure used in systems (a step that will require cooperation with commercial providers and civilian authorities).

Few networks are built entirely using systems controlled by the organization that relies on them. Therefore organizations (including DOD) are required to work cooperatively with the owners of the infrastructure they rely on and relevant authorities to protect them.

Function 8. Include security requirements in any specification of system or network requirements that is used in the acquisition process.

Providing information systems security for a network or system that has not had security features built into it is enormously problematic. Retrofits of security features into systems not designed for security invariably leave security holes, and procedural fixes for inherent technical vulnerabilities only go so far.

Perhaps more importantly, security requirements must be given prominence from the beginning in any system conceptualization. The reason is that security considerations may affect the design of a system in quite fundamental ways, and a designer who decides on a design that works against security should at least be cognizant of the implications of such a choice. This function thus calls for information systems security expertise to be integrally represented on design teams, rather than added later.

Note that specification of the "Orange Book" security criteria would be an insufficient response to this function. "Orange Book" criteria typically drive up development times significantly, and more importantly, are not inherently part of an initial requirements process and do not address the security of networked or distributed systems.

Function 9. Monitor, assess, and understand offensive and defensive information technologies.

Good information systems security requires an understanding of the types of threats and defenses that might be relevant. Thus, those responsible for information systems security need a vigorous ongoing program to monitor, assess, and understand offensive and defensive information technologies. Such a program would address the technical details of these technologies, their capability to threaten or protect friendly systems, and their availability.

Function 10. Advance the state of the art in defensive information technology (and processes) with research.

Although much can usually be done to improve information systems security simply through the use of known and available technologies, "bug fixes," and procedures, better tools to support the information systems security mission are always needed. In general, such improvements fall into two classes (which may overlap). One class consists of improvements so that tools can deal more effectively with a broader threat spectrum. A second class, equally important, provides tools that provide better automation and thus can solve problems at lower costs (costs that include direct outlays for personnel and equipment and operational burdens resulting from the hassle imposed by providing security).

Similar considerations apply to processes for security as well. It is reasonable to conduct organizational research into better processes and organizations that provide more effective support against information attacks and/or reduce the impediments to using or implementing good security practices.

Function 11. Promote information systems security awareness.

Just as it is dangerous to rely on a defensive system or network architecture that is hard on the outside and soft on the inside, it is also dangerous if any member of an organization fails to take information systems security seriously. Because the carelessness of a single individual can seriously compromise the security of an entire organization, education and training for information systems security must be required for all members of the organization. Moreover, such education and training must be systematic, regarded as important by the organization (and demonstrated with proper support for such education and training), and undertaken on a regular basis (both to remind people of its importance and to update their knowledge in light of new developments in the area).

Function 12. Set security standards (both technical and procedural).

Security standards should articulate in well-defined and actionable terms what an organization expects to do in the area of security. They are therefore prescriptive. For example, a technical standard might be "all passwords must be eight or more characters long, contain both letters and numbers, be pronounceable, and not be contained in any dictionary," or "all electronic communications containing classified information must be encrypted with a certain algorithm and key length." A standard involving both technical and procedural measures might specify how to revoke cryptographic keys known to have been compromised. Furthermore, security standards should be expected to apply to all those within the organization. (For example, generals should not be allowed to exercise poorer information systems security discipline than do captains, as they might be tempted to do in order to make their use of C4I systems easier.)

Function 13. Develop and use criteria for assessing security status.

Information security is not a one-shot problem, but a continuing one. Threats, technology, and organizations are constantly changing in a spiral of measures and countermeasures. Organizations must have ways of measuring and evaluating whether they have effective defensive measures in place. Thus, once standards are put in place, the organization must periodically assess the extent to which members of the organization comply with those standards, and characterize the nature of the compliance that does exist.

Metrics for security could include number of attacks of different types, fraction of attacks detected, fraction of attacks repelled, damage incurred, and time needed to detect and respond to attacks. Note that making measurements on such parameters depends on understanding the attacks that do occur—because many attacks are not detected today, continual penetration testing is required to establish such a baseline.

On Practice in the Field

(From pp. 156-157): . . .The security in today's fielded military systems is weak, and weaker than it need be, as illustrated by the following examples of behavior and practices that the committee observed or heard:

• Individual nodes are running commercial software with many known security problems. Operators use little in the way of tools for finding these problems, to say nothing of fixing them.
• Computers attached to sensitive command and control systems are also used by personnel to surf Web sites worldwide, raising the possibility that rogue applets and the like could be introduced into the system.
• Units are being blinded by denial-of-service attacks, made possible because individual nodes were running commercial software with many known security problems.
• IP addresses and other important data about C2 [command and control] systems can be found on POST-IT notes attached to computers in unsecured areas, making denial of service and other attacks much easier.
• Some of the networks used by DOD to carry classified information are protected by a perimeter defense. As a result, they exhibit all of the

vulnerabilities that characterize networks protected by perimeter defenses.

(From p. 158): Many field commanders told the committee that "cyberspace is part of the battlespace," and several organizations within the DOD assert that they are training "C2/cyber warriors." But good intentions have not been matched by serious attention to cyberspace protection. Soldiers in the field do not take the protection of their C4I systems nearly as seriously as they do other aspects of defense. For example, information attack red teams were a part of some exercises observed by the committee, but their efforts were usually highly constrained for fear that unconstrained efforts would bring the exercise to a complete halt. While all red teams operate under certain rules of engagement established by the "white team" that oversees each exercise, the information attack red teams appeared to the committee to be much more constrained than appropriate. In one exercise, personnel in an operations center laughed and mistakenly took as a joke a graphic demonstration by the red team that their operations center systems had been penetrated.

TRUST IN CYBERSPACE (1999)

CITATION: Computer Science and Telecommunications Board (CSTB), National Research Council. 1999. *Trust in Cyberspace.* National Academy Press, Washington, D.C.

Cybersecurity and Other Trustworthiness Qualities Interact

(From p. 14): The trustworthiness of [a networked information system] encompasses correctness, reliability, security (conventionally including secrecy, confidentiality, integrity, and availability), privacy, safety, and survivability . . . . These dimensions are not independent, and care must be taken so that one is not obtained at the expense of another. For example, protection of confidentiality or integrity by denying all access trades one aspect of security—availability—for others. As another example, replication of components enhances reliability but may increase exposure to attack owing to the larger number of sites and the vulnerabilities implicit in the protocols to coordinate them. Integrating the diverse dimensions of trustworthiness and understanding how they interact are central challenges in building a trustworthy [networked information system].

On Managing Risk

(From pp. 175-176): A discussion about consequences must also address the questions of who is affected by the consequences and to what extent. While catastrophic failure garners the most popular attention, there are many dimensions to trustworthiness and consequences may involve various subsets of them with varying degrees of severity. . . . Understanding consequences is essential to forming baseline expectations of private action and what incentives may be effective for changing private action, but that understanding is often hampered by the difficulty of quantifying or otherwise specifying the costs and consequences associated with risks.

(From p. 175): It is the nature of [a networked information system] that outages and disruptions of service in local areas may have very uneven consequences, even within the area of disruption. Failure of a single Internet service provider (ISP) may or may not affect transfer of information outside the area of disruption, depending on how the ISP has configured its communications. For example, caching practices intended to reduce network congestion problems helped to limit the scope of a Domain Name Service (DNS) outage. Corporations that manage their own interconnection (so-called intranets) may be wholly unaffected. Even widespread or catastrophic failures may not harm some users, if they have intentionally or unconsciously provided redundant storage or backup facilities. The inability to accurately predict consequences seriously complicates the process of calculating risk and makes it tempting to assume "best case" behavior in response to failure.

(From pp. 177-178): . . . [T]he costs associated with avoiding all risks are prohibitive. Thus, risk mitigation is more typical and is generally encountered when many factors, including security and reliability, determine the success of a system. Risk mitigation is especially popular in market-driven environments where an attempt is made to provide "good enough" security or reliability or other qualities without severely affecting economic factors such as price and time to market. Risk mitigation should be interpreted not as a license to do a shoddy job in implementing trustworthiness, but instead as a pragmatic recognition that trade-offs between the dimensions of trustworthiness, economic realities, and other constraints will be the norm, not the exception. The risk mitigation strategies that are most relevant to trustworthiness can generally be characterized according to two similar models:

- *The insurance model.* In this model, the cost of countermeasures is viewed as an "insurance premium" paid to prevent (or at least mitigate) loss. The value of the information being protected, or the service being provided, is assessed and mechanisms and assurance steps are incorporated up to, but not exceeding, that value.
- *The work factor model.* A definition in cryptology for the term "work factor" is the amount of computation required to break a cipher through a brute-force search of all possible key values. Recently, the term has been broadened to mean the amount of effort required to locate and exploit a residual vulnerability. That effort may involve more efficient procedures rather than exhaustive searches. In the case of fault tolerance, the assumptions made about the types of failures (benign or arbitrary) that could arise are analogous to the concept of work factor.

The two models are subject to pitfalls distinctive to each and some that are common to both. In the insurance model, it is possible that the value of information (or disruption of service) to an outsider is substantially greater than the value of that information or service to its owners. Thus, a "high value" attack could be mounted, succeed, and the "insurance premium" lost along with the target data or service. Such circumstances often arise in an interconnected or networked world. For example, a local telephone switch might be protected against deliberate interruption of service to the degree that is justified by the revenue that might be lost from such an interruption. But such an analysis ignores the attacker whose aim is to prevent a physical alarm system from notifying the police that an intrusion has been detected into an area containing valuable items. Another example is an instance in which a hacker expends great effort to take over an innocuous machine, not because it contains interesting data but because it provides computing resources and network connectivity that can be used to mount attacks on higher-value targets. In the case of the work factor model, it is notoriously difficult to assess the capabilities of a potential adversary in a field as unstructured as that of discovering vulnerabilities, which involves seeing aspects of a system that were overlooked by its designers.

Vulnerabilities in the Public Telephone Network and the Internet

(From p. 27): The vulnerabilities of the PTN [public telephone network] and Internet are exacerbated by the dependence of each network on the other. Much of the Internet uses leased telephone lines as its physical transport medium. Conversely, telephone companies rely on networked computers to manage their own facilities, increasingly employing Internet technology, although not necessarily the Internet itself. Thus, vulnerabilities in the PTN can affect the Internet, and vulnerabilities in Internet technology can affect the telephone network.

(From p. 58): . . . [W]hile in one sense the Internet poses no new challenges—a system that can be accessed from outside only through a cryptographically protected channel on the Internet is at least as secure as the same system reached through a conventional leased line—new dangers arise precisely because of pervasive interconnectivity. The capability to interconnect networks gives the Internet much of its power; by the same token, it opens up serious new risks. An attacker who may be deflected by cryptographic protection of the front door can often attack a less protected administrative system and use its connectivity through internal networks to bypass the encryption unit protecting the real target. This often makes a mockery of firewall-based protection.

(From p. 50): The general accessibility of the Internet makes it a highly visible target and within easy reach of attackers. The widespread availability of documentation and actual implementations for Internet protocols means that devising attacks for this system can be viewed as an intellectual puzzle (where launching the attacks validates the puzzle's solution).

CYBERSECURITY TODAY AND TOMORROW

On Building Secure Systems and Networks

(From p. 2): Laudable as a goal, ab initio building of trustworthiness into an NIS [networked information system] has proved to be impractical. It is neither technically nor economically feasible for designers and builders to manage the complexity of such large artifacts or to anticipate all of the problems that an NIS will confront over its lifetime. Experts now recognize steps that can be taken to enhance trustworthiness after a system has been deployed. It is no accident that the market for virus detectors and firewalls is thriving. Virus detectors identify and eradicate attacks embedded in exchanged files, and firewalls hinder attacks by filtering messages between a trusted enclave of networked computers and its environment (from which attacks might originate). Both of these mechanisms work in specific contexts and address problems contemplated by their designers; but both are imperfect, with user expectations often exceeding what is prudent.

(From pp. 13-14): Networked information systems (NISs) integrate computing systems, communications systems, and people (both as users and operators). The defining elements are interfaces to other systems along with algorithms to coordinate those systems. Economics dictates the use of commercial off-the-shelf (COTS) components wherever possible, which means that developers of an NIS have neither control over nor detailed information about many system components. The use of system components whose functionality can be changed remotely and while the system is running is increasing. Users and designers of an NIS built from such extensible system components thus cannot know with any certainty what software has entered system components or what actions those components might take.

(From p. 3): Today's climate of deregulation will further increase [networked information system] vulnerability in several ways. The most obvious is the new cost pressures on what had been regulated monopolies in the electric power and telecommunications industries. One easy way to cut costs is to reduce reserve capacity and eliminate rarely needed emergency systems; a related way is to reduce diversity (a potential contributor to trustworthiness) in the technology or facilities used. Producers in these sectors are now competing on the basis of features, too. New features invariably lead to more complex systems, which are liable to behave in unexpected and undesirable ways. Finally, deregulation leads to new interconnections, as some services are more cost-effectively imported from other providers into what once were monolithic systems. Apart from the obvious dangers of the increased complexity, the interconnections themselves create new weak points and interdependencies. Problems could grow beyond the annoyance level that characterizes infrastructure outages today, and the possibility of catastrophic incidents is growing.

(From p. 15): To be labeled as trustworthy, a system not only must behave as expected but also must reinforce the belief that it will continue to produce expected behavior and will not be susceptible to subversion. The question of how to achieve assurance has been the target of several research programs sponsored by the Department of Defense and others. Yet currently practiced and proposed approaches for establishing assurance are still imperfect and/or impractical. Testing can demonstrate only that a flaw exists, not that all flaws have been found; deductive and analytical methods are practical only for certain small systems or specific properties. Moreover, all existing assurance methods are predicated on an unrealistic assumption—that system designers and implementors know what it means for a system to be "correct" before and during development. The study committee believes that progress in assurance for the foreseeable future will most likely come from figuring out (1) how to combine multiple approaches and (2) how best to leverage add-on technologies and other approaches to enhance existing imperfect systems. Improved assurance, without any pretense of establishing a certain or a quantifiable level of assurance, should be the aim.

(From p. 247): Security research during the past few decades has been based on formal policy models that focus on protecting information from unauthorized access by specifying which users should have access to data or other system objects. It is time to challenge this paradigm of "absolute security" and move toward a model built on three axioms of insecurity: insecurity exists; insecurity cannot be destroyed; and insecurity can be moved around.

(From p. 250): Improved trustworthiness may be achieved by the careful organization of untrustworthy components. There are a number of promising ideas, but few have been vigorously pursued. "Trustworthiness from untrustworthy components" is a research area that deserves greater attention.

On the Impact of System Homogeneity ("Monoculture")

(From pp. 191-192): The similarity intrinsic in the component systems of a homogeneous collection implies that these component systems share vulnerabilities. A successful attack on one system is then likely to succeed on other systems as well—the antithesis of what is desired for implementing trustworthiness. Moreover, today's dominant computing and communications environments are based on hardware and software that were not designed with security in mind; consequently, these systems are not difficult to compromise, as discussed in previous chapters.

There is, therefore, some tension between homogeneity and trustworthiness. Powerful forces make technological homogeneity compelling . . ., but some attributes of trustworthiness benefit from diversity. . . . On the other hand, a widely used trustworthy operating system might be superior to a variety of nontrustworthy operating systems; diversity, per se, is not equivalent to increased trustworthiness.

Technological convergence may also be realized through the market dominance of a few suppliers of key components, with monopoly as the limit case when technological homogeneity is dictated by the monopolist. However, the number of suppliers could grow as a result of the diffusion of computing into embedded, ubiquitous environments; the diversification and interoperability of communications services; and the continued integration of computing and communications into organizations within various market niches.

What Is CSTB?

As a part of the National Research Council, the Computer Science and Telecommunications Board (CSTB) was established in 1986 to provide independent advice to the federal government on technical and public policy issues relating to computing and communications. Composed of leaders from industry and academia, CSTB conducts studies of critical national issues and makes recommendations to government, industry, and academic researchers. CSTB also provides a neutral meeting ground for consideration of complex issues where resolution and action may be premature. It convenes invitational discussions that bring together principals from the public and private sectors, ensuring consideration of all perspectives. The majority of CSTB's work is requested by federal agencies and Congress, consistent with its National Academies context.

A pioneer in framing and analyzing Internet policy issues, CSTB is unique in its comprehensive scope and effective, interdisciplinary appraisal of technical, economic, social, and policy issues. Beginning with early work in computer and communications security, cyber-assurance and information systems trustworthiness have been a cross-cutting theme in CSTB's work. CSTB has produced several reports regarded as classics in the field, and it continues to address these topics as they grow in importance.

To do its work, CSTB draws on some of the best minds in the country, inviting experts to participate in its projects as a public service. Studies are conducted by balanced committees without direct financial interests in the topics they are addressing. Those committees meet, confer electronically, and build analyses through their deliberations. Additional expertise from around the country is tapped in a rigorous process of review and critique, further enhancing the quality of CSTB reports. By engaging groups of principals, CSTB obtains the facts and insights critical to assessing key issues.

The mission of CSTB is to:

- *Respond to requests* from the government, nonprofit organizations, and private industry for advice on computer and telecommunications issues and from the government for advice on computer and telecommunications systems planning, utilization, and modernization;
- *Monitor and promote the health of the fields* of computer science and telecommunications, with attention to issues of human resources, information infrastructure, and societal impacts;
- *Initiate and conduct studies* involving computer science, computer technology, and telecommunications as critical resources; and
- *Foster interaction* among the disciplines underlying computing and telecommunications technologies and other fields, at large and within the National Academies.

As of February 2002, current CSTB activities with a cybersecurity component address privacy in the information age, critical information infrastructure protection, authentication technologies and their privacy implications, information technology for counteracting terrorism, and geospatial information systems. Additional studies examine digital government, the fundamentals of computer science, limiting children's access to pornography on the Internet, information technology and creativity, computing and biology, and Internet navigation and the Domain Name System. Explorations are under way in the areas of the insider threat, dependable and safe software systems, wireless communications and spectrum management, digital archiving and preservation, open source software, digital democracy, the "digital divide," manageable systems, information technology and journalism, and women in computer science.

More information about CSTB can be obtained online at <http://www.cstb.org>.